Neptune

The Whisper of Dreams

JD ARDEN

Neptune

Preface: The Distant Guardian

Neptune, the eighth and outermost planet in the solar system, occupies a liminal space between the known and the unknown. It is the last of the gas giants, a distant sentinel standing on the threshold of the Kuiper Belt and the vast, uncharted realms beyond. Its vivid blue hue, turbulent atmosphere, and dynamic moons make it a world of striking beauty and profound mystery. Neptune's remoteness, both literal and metaphorical, invites reflection on the nature of boundaries—between light and shadow, warmth and cold, presence and absence.

Discovered through the power of mathematics rather than direct observation, Neptune is a testament to human ingenuity and the ability to perceive what lies beyond the reach of the senses. Its discovery in the mid-19th century marked a milestone in astronomy, revealing a universe governed not only by chance but by patterns and predictability. Yet Neptune itself remains enigmatic, its secrets veiled by distance and time.

This is a planet of contradictions: stormy yet serene, cold yet luminous, distant yet deeply connected to the forces that shaped the solar system. Its Great Dark Spot, a transient storm larger than Earth, highlights the dynamic energy coursing through its frigid atmosphere. Its largest moon, Triton, with its retrograde orbit and icy plumes, suggests a complex history tied to the Kuiper Belt and beyond. Even Neptune's faint rings, ephemeral and clumpy, whisper of processes that challenge our understanding of the outer solar system.

Neptune's place in mythology, as the Roman god of the sea, mirrors its character as a world of depths and currents, both physical and symbolic. It is a planet that invites us to dream of distant horizons and to confront the limits of our knowledge. Neptune is not just a world to be studied; it is a world to be felt, a reminder of the vastness of the cosmos and the enduring human drive to explore its furthest reaches.

In this book, Neptune emerges as both a scientific marvel and a philosophical muse. From the fastest winds in the solar system to its role as a guardian of the outer solar system, Neptune challenges us to think expansively about what lies beyond. Its story is one of connection, between past and future, near and far, known and unknown. Neptune is

Neptune

not merely a boundary; it is a bridge—a whisper of dreams that calls us to venture further into the cosmic depths.

Chapter 1: Neptune's Turbulent Atmosphere

Neptune's atmosphere is a realm of extremes, where the fastest winds in the solar system race through a cold, distant world. At first glance, it may seem paradoxical that such ferocious storms occur on a planet more than 30 times farther from the Sun than Earth, in a region where solar energy is a mere fraction of what it is in the inner solar system. Yet, Neptune's atmosphere is anything but static. It is a dynamic, churning system, driven by forces that continue to defy expectations and inspire inquiry.

The most iconic feature of Neptune's atmosphere is the Great Dark Spot, a massive storm system comparable in size to Earth. First observed by the Voyager 2 spacecraft during its flyby in 1989, the Great Dark Spot was a transient phenomenon, disappearing within a few years but leaving a lasting impression on planetary science. Unlike Jupiter's Great Red Spot, which has persisted for centuries, Neptune's storms are fleeting, appearing and dissipating over relatively short periods. These storms are markers of Neptune's dynamic energy, highlighting the complex interplay of atmospheric currents, temperatures, and internal heat.

Beneath Neptune's pale blue clouds lies a layered atmosphere dominated by hydrogen and helium, with traces of methane that give the planet its vibrant color. Methane absorbs red light and reflects blue, creating a hue that is strikingly different from the muted tones of Uranus. However, methane alone cannot account for Neptune's deep azure appearance. Other, less understood factors—perhaps high-altitude hazes or the scattering of sunlight by unseen compounds—contribute to the planet's vivid coloration.

The winds of Neptune are a defining characteristic, reaching speeds of up to 2,100 kilometers per hour (about 1,300 miles per hour). These are the fastest winds measured in the solar system, yet their origins remain a subject of debate. One contributing factor is Neptune's internal heat, which radiates more energy into space than it receives from the Sun.

Neptune

This heat likely drives convection currents within the atmosphere, powering the planet's extreme weather.

The cold temperatures of Neptune's upper atmosphere, which can drop as low as -220°C (-364°F), create sharp contrasts with the warmer layers below. This temperature gradient fuels turbulence, as warmer gases rise and cooler ones sink, generating powerful jet streams and storms. The rapid rotation of Neptune, which completes a day in just 16 hours, further accelerates these winds, creating a banded structure of atmospheric currents that spiral across the planet.

Despite these insights, Neptune's storms remain a puzzle. Observations from Earth-based telescopes and the Hubble Space Telescope have revealed bright clouds and dark vortices forming and disappearing over time, suggesting a dynamic and ever-changing system. These storms are not merely random; they appear to be shaped by underlying structures within the planet's atmosphere, as well as by interactions with Neptune's magnetic field and possibly its internal dynamics.

The Great Dark Spot and its successors are more than meteorological phenomena; they are windows into the forces that drive Neptune's atmosphere. Their transient nature contrasts with the stability of Jupiter's Great Red Spot, highlighting the diversity of storm systems in the solar system. These storms also raise questions about the energy sources and feedback mechanisms that sustain Neptune's atmosphere, offering clues to the behavior of other gas giants and even exoplanets with similar compositions.

Neptune's atmosphere, with its extremes of wind, temperature, and turbulence, challenges traditional models of planetary weather. It forces scientists to reconsider the factors that shape atmospheric dynamics, from internal heat and rotation to the influence of magnetic fields and solar radiation. Neptune serves as a natural laboratory for studying these processes, extending our understanding of how atmospheres function not only in our solar system but also on distant worlds.

Philosophically, Neptune's turbulent atmosphere invites reflection on the paradox of calm within chaos. Beneath its serene blue exterior lies a world of motion and energy, a reminder that appearances can be deceiving. The planet's winds, storms, and transient features are a

Neptune

testament to the complexity of nature, where forces operate on scales and in ways that often elude human comprehension.

To gaze upon Neptune is to witness a dance of extremes—a world where the coldest outer layers conceal the hottest inner energies, and where the fastest winds emerge in the most remote and frigid regions of the solar system. Neptune's atmosphere is a celebration of contrast and motion, a symbol of the universe's boundless creativity and the forces that shape its most distant corners.

Chapter 2: A Luminous Blue Mystery

Neptune's striking blue color, a vivid and almost otherworldly hue, sets it apart from every other planet in the solar system. While its neighbor Uranus also appears blue due to the presence of methane in its atmosphere, Neptune's blue is richer and deeper, a visual signature that hints at complexities beneath its clouds. This color is more than an aesthetic feature; it is a clue to the processes that govern Neptune's atmosphere, the interplay of light, heat, and chemistry, and the mysterious forces that sustain its luminous beauty.

At the heart of Neptune's blue lies **methane**, a compound present in trace amounts in its upper atmosphere. Methane absorbs red wavelengths of sunlight and reflects blue, giving the planet its basic coloration. This process is well understood and explains why both Neptune and Uranus exhibit blue hues. However, methane alone cannot account for the stark difference in the intensity of their colors. Uranus's pale aquamarine is muted compared to Neptune's deep azure, suggesting that additional factors are at play.

One possibility lies in the **vertical structure** of Neptune's atmosphere. Unlike Uranus, which has minimal internal heat, Neptune radiates more energy than it receives from the Sun. This heat likely drives convection currents, mixing gases and aerosols throughout the atmosphere. The turbulent motion could create variations in the distribution of haze and clouds, affecting how light interacts with different layers. In this sense, Neptune's deeper blue might reflect a more active and dynamic atmosphere.

High-altitude hazes, composed of tiny particles or aerosols, are another key factor. These hazes scatter sunlight differently depending on their size, composition, and altitude. On Neptune, the precise makeup of these particles remains uncertain, but they likely include complex hydrocarbons formed by the interaction of sunlight and methane. These hydrocarbons may absorb certain wavelengths of light, enhancing the planet's vivid blue by suppressing other colors.

The interplay between **light scattering and absorption** also contributes to Neptune's unique coloration. Rayleigh scattering, the same

Neptune

phenomenon that makes Earth's sky appear blue, is at work in Neptune's upper atmosphere. However, on Neptune, this scattering occurs alongside absorption by methane and possibly other compounds, creating a more intense and uniform blue. The exact balance of these processes remains a subject of ongoing research, highlighting the complexity of planetary atmospheres and the challenge of unraveling their secrets.

Neptune's blue hue is further enhanced by its distance from the Sun. At such a great distance, sunlight is faint, and the energy available for driving atmospheric reactions is minimal. Yet, despite this lack of solar energy, Neptune's atmosphere remains active, with bright clouds and dynamic weather systems forming against its vivid backdrop. This apparent contradiction underscores the role of **internal heat** in sustaining the planet's atmospheric processes.

Neptune's internal heat is a defining characteristic, setting it apart not only from Uranus but also from many other planets. The source of this heat is thought to lie in the planet's deep interior, where gravitational compression, radioactive decay, and other processes generate energy. This heat rises through Neptune's layers, fueling convection currents and driving the formation of storms, jet streams, and cloud bands. The interaction of this internal energy with the cold outer atmosphere creates temperature gradients that shape Neptune's weather and contribute to its striking appearance.

Interestingly, Neptune's blue color is not static. Observations over time have revealed subtle changes in its hue, likely tied to seasonal variations and atmospheric dynamics. As Neptune orbits the Sun over its 165-year journey, the distribution of sunlight across its atmosphere shifts, affecting the behavior of clouds, hazes, and winds. These changes are slow and difficult to observe from Earth, but they suggest a planet in flux, where even its iconic blue evolves over time.

The color of Neptune has implications beyond its aesthetic appeal. It offers a window into the processes that shape not only Neptune but also other ice giants and exoplanets. Many exoplanets discovered to date are similar in size and composition to Neptune, and their colors, spectra, and atmospheric dynamics can provide clues about their habitability, weather systems, and evolutionary histories. By studying Neptune's blue,

Neptune

scientists refine their models of atmospheric physics and extend those insights to the wider universe.

Philosophically, Neptune's blue is a reminder of the interplay between simplicity and complexity. At first glance, its color might seem like a straightforward result of methane's presence, but deeper exploration reveals a tapestry of processes—scattering, absorption, turbulence, and heat transfer—that combine to create its luminous appearance. Neptune's blue invites us to look beyond the surface, to consider the hidden forces that shape even the most apparent features of a planet.

The vibrant blue of Neptune also speaks to the power of light and perception. It is a color that exists not as a tangible object but as an interaction between sunlight, molecules, and the human eye. In this way, Neptune's blue is both a physical phenomenon and a subjective experience, a fusion of cosmic forces and human observation.

To gaze upon Neptune is to witness a world that defies its distance. Its color is not merely a reflection of sunlight but a beacon that connects it to the rest of the solar system and beyond. Neptune's blue is a bridge between the familiar and the unknown, a reminder that even in the coldest, most remote corners of the cosmos, beauty thrives.

Chapter 3: Triton's Secrets

Among Neptune's moons, **Triton** stands apart as a world of singular intrigue. It is not merely Neptune's largest satellite; it is one of the most fascinating objects in the solar system. Triton's unique characteristics—a retrograde orbit, icy geysers, and a surface shaped by cryovolcanism—make it a dynamic and enigmatic world that continues to captivate scientists and dreamers alike. It is a moon that tells a story not only of Neptune's system but also of the far reaches of the Kuiper Belt and the processes that govern planetary formation and evolution.

Triton's most striking feature is its **retrograde orbit**, meaning it moves in the opposite direction of Neptune's rotation. This sets it apart from most large moons in the solar system, which typically follow prograde orbits aligned with their planet's spin. Triton's retrograde motion strongly suggests that it was not formed alongside Neptune but was instead a captured object, likely originating from the Kuiper Belt—a region of icy bodies beyond Neptune's orbit.

The process by which Triton was captured remains a topic of scientific investigation. One leading hypothesis is that Triton was once part of a binary system, where two Kuiper Belt objects orbited each other. As this pair passed near Neptune, gravitational interactions disrupted their balance, causing one object to be ejected while the other, Triton, was captured by Neptune. This event would have been catastrophic, destabilizing any existing moons of Neptune and reshaping the planet's satellite system. The aftermath of Triton's capture is evident in its unusual orbit, which is both retrograde and highly inclined, marking it as a relic of a dramatic past.

Triton's surface is a patchwork of icy plains, ridges, and depressions, shaped by processes that are rare or nonexistent on other moons. One of the most remarkable features of Triton's landscape is its evidence of **cryovolcanism**—a phenomenon where icy material, rather than molten rock, erupts from below the surface. Voyager 2, during its flyby in 1989, captured images of active geysers on Triton, ejecting plumes of nitrogen gas and dust several kilometers into space. These geysers are thought to be driven by solar heating, where sunlight penetrates a thin layer of surface ice, vaporizing subsurface nitrogen and causing it to erupt.

The presence of active geysers and cryovolcanism on Triton points to a dynamic interior. Despite its distance from the Sun and its icy composition, Triton appears to retain enough internal heat to drive geological activity. This heat could originate from tidal interactions with Neptune, particularly in the past, when Triton's orbit was more eccentric and the tidal forces stronger. Another potential source of heat is radioactive decay within the moon's rocky core, providing a steady but faint energy source.

Triton's icy surface is among the coldest in the solar system, with temperatures plummeting to around -235°C (-391°F). Despite this frigid environment, the surface is remarkably smooth, with few impact craters. This suggests that Triton's surface is relatively young, constantly reshaped by geological and cryovolcanic processes. Vast plains of nitrogen ice, interspersed with ridges and fractures, create a landscape that is both alien and dynamic, a reminder of the diversity of worlds that exist even in the outermost reaches of the solar system.

One of the most intriguing questions about Triton is its potential to host a **subsurface ocean**. The combination of internal heat, a rocky core, and an icy shell creates conditions that might allow liquid water to exist beneath its frozen crust. If such an ocean exists, it could be a site for prebiotic or even biological activity, akin to the subsurface oceans suspected on Europa and Enceladus. While the extreme cold and lack of sunlight would make life as we know it unlikely, the presence of organic compounds and energy sources raises the possibility of alternative chemistries capable of supporting life.

Triton's role within Neptune's system is also significant. Its massive size and retrograde orbit affect Neptune's other moons and rings, creating gravitational interactions that shape their orbits and dynamics. Triton is slowly spiraling inward toward Neptune due to tidal forces, a process that will eventually result in its destruction. In several billion years, Triton will cross Neptune's Roche limit, where the planet's gravity will tear the moon apart, potentially forming a new ring system around Neptune. This distant but inevitable event underscores the transient nature of even the most stable-seeming celestial bodies.

Neptune

Voyager 2's brief visit to Triton provided humanity's first and only close-up views of this enigmatic moon. The images and data it returned transformed Triton from a distant point of light into a world of active geology, icy landscapes, and potential habitability. Yet, much remains unknown. What drives its internal heat? How thick is its icy crust? Could a subsurface ocean exist, and if so, what might it contain? These questions await the arrival of future missions, which could probe Triton's surface, atmosphere, and interior with advanced instruments.

Philosophically, Triton serves as a reminder of the interconnectedness of the solar system. Its origins in the Kuiper Belt link it to a vast and ancient region of icy bodies, remnants from the early solar system's formation. Its capture by Neptune reshaped both itself and the planet it now orbits, highlighting the dynamic interactions that shape celestial systems. Triton is a moon that defies isolation; its story is one of movement, collision, and transformation, a narrative that connects it to the broader history of the solar system.

Triton's secrets are not merely scientific; they are also symbolic. Its active geysers, retrograde orbit, and icy plains challenge conventional notions of moons as static and inert objects. Triton reminds us that even in the coldest and most distant corners of the solar system, there is dynamism, creativity, and the potential for discovery. It is a world that whispers of the unknown, inviting us to explore further and dream bigger.

As we gaze toward Neptune and its enigmatic moon, Triton becomes a symbol of possibility—a reminder that the most profound discoveries often lie in the most unexpected places. Its secrets are a call to curiosity, a testament to the power of exploration, and an invitation to unravel the mysteries of a world unlike any other.

Chapter 4: Neptune's Rings

The rings of Neptune, faint and delicate compared to the luminous bands of Saturn, are a testament to the diversity and complexity of planetary ring systems in the solar system. Discovered in the late 20th century and studied in detail during the **Voyager 2 flyby**, these rings defy expectations with their clumpiness, narrow structure, and transient nature. They are both a scientific curiosity and a symbol of Neptune's enigmatic character—a quiet yet intricate feature that reflects the forces shaping the outer solar system.

Neptune's rings were first suspected during Earth-based observations in the 1980s, when astronomers noticed a dimming of starlight as it passed near Neptune, suggesting the presence of a ring system. However, these observations were inconsistent, leading to speculation that Neptune might have incomplete or partial rings—arc-like structures rather than fully connected bands. This mystery was resolved in 1989 when Voyager 2 revealed that Neptune indeed had rings, but they were faint, unevenly distributed, and far less conspicuous than those of Saturn or even Uranus.

Neptune's ring system consists of five principal rings, named **Galle**, **Le Verrier**, **Lassell**, **Arago**, and **Adams**, after astronomers who contributed to Neptune's discovery and study. These rings lie within the planet's Roche limit, the region where tidal forces prevent material from coalescing into larger moons. Each ring has unique characteristics, shaped by a combination of gravitational interactions, collisions, and processes that remain only partially understood.

The **Adams ring**, the outermost of Neptune's rings, is perhaps the most intriguing. It contains several **arcs**, dense clumps of material that defy the expected uniform distribution of particles within a ring. These arcs—given names like Liberty, Equality, and Fraternity—are stable formations that remain confined despite the forces that should disperse them. The precise mechanism maintaining these arcs is not fully understood, but their stability is thought to be influenced by the gravitational effects of nearby moons, particularly **Galatea**, a small moon that orbits just inside the Adams ring.

Neptune

The clumpiness of the Adams ring and its arcs raises questions about the dynamics of ring systems. Unlike the broad, uniform rings of Saturn, Neptune's rings are highly uneven, with variations in density and brightness that suggest an ongoing process of formation and dissipation. These variations are likely the result of collisions between ring particles, interactions with Neptune's moons, and the influence of the planet's magnetic field.

The other rings of Neptune, while less dramatic, are equally fascinating. The **Le Verrier** and **Galle** rings are narrow and faint, composed of fine dust and ice particles that scatter light weakly, making them difficult to observe even with advanced instruments. The **Lassell** and **Arago** rings, located closer to Neptune, are even more diffuse, appearing as faint bands of material that blend into the planet's hazy surroundings. These inner rings are thought to be shaped by similar processes to those affecting the outer rings, including the gravitational influence of small, unseen moons.

The composition of Neptune's rings adds another layer of complexity. They are believed to be made of dark material, possibly organic compounds mixed with silicate and icy particles. This composition gives the rings their low albedo, or reflectivity, making them much darker than Saturn's bright, icy rings. The source of this material is likely a combination of dust from micrometeoroid impacts, remnants of ancient moons that were shattered by collisions, and material ejected from Neptune's smaller moons.

One of the most intriguing aspects of Neptune's rings is their **ephemeral nature**. Unlike Saturn's rings, which appear relatively stable over geological timescales, Neptune's rings are thought to be transitory, constantly evolving and potentially short-lived in cosmic terms. The faint, diffuse bands and uneven clumps suggest a system in flux, where material is added and lost over time. This transience raises questions about the origins of the rings and their eventual fate.

The interaction between Neptune's rings and its moons highlights the interconnectedness of the Neptune system. Small moons like Galatea act as **shepherds**, their gravity confining ring particles and maintaining the narrow structure of certain rings. At the same time, the rings contribute material to the surfaces of these moons, coating them with dust and ice.

Neptune

This dynamic relationship underscores the delicate balance of forces that shape Neptune's rings and moons, creating a system that is both intricate and fragile.

The study of Neptune's rings also provides broader insights into the processes that govern planetary ring systems across the solar system. By comparing Neptune's rings to those of Saturn, Uranus, and Jupiter, scientists can identify commonalities and differences that shed light on the factors influencing ring formation, evolution, and stability. Neptune's rings, with their clumpiness and transience, represent a unique case that challenges existing models and expands our understanding of how rings behave in diverse environments.

Philosophically, Neptune's rings are a reminder of the beauty in impermanence. Their faintness, clumpiness, and transitory nature contrast with the grandeur of Saturn's rings, yet they possess a quiet elegance that rewards close observation. They speak to the dynamic processes that create and sustain celestial structures, even in the cold, remote reaches of the solar system. Neptune's rings invite us to reflect on the fragility of cosmic systems and the fleeting nature of even the most intricate formations.

Neptune's rings, though subtle and understated, are a vital part of its story. They are not merely decorative features but dynamic systems that reflect the planet's history, interactions, and ongoing evolution. To study these rings is to engage with the processes that shape not only Neptune but the solar system as a whole, offering a window into the forces that govern the universe.

In their clumps and gaps, their faint arcs and diffuse bands, Neptune's rings whisper of a world in motion—a world where beauty emerges not from permanence but from the delicate balance of forces that sustain it. They are a symbol of Neptune's quiet complexity and a testament to the diversity of the outer solar system.

Chapter 5: Discovery Through Mathematics

Neptune's discovery is a story of human ingenuity, perseverance, and the power of mathematics to uncover truths that lie beyond the reach of the senses. Unlike the other planets visible to the naked eye since antiquity, Neptune was the first planet to be discovered not through direct observation but through calculations based on gravitational theory. This remarkable achievement, occurring in the mid-19th century, marked a turning point in astronomy, demonstrating that the universe operates according to discernible patterns and principles that can be understood and predicted through the application of reason and science.

The story of Neptune's discovery begins with its predecessor in the solar system, Uranus. Discovered by William Herschel in 1781, Uranus expanded the known boundaries of the solar system and brought with it a new challenge for astronomers: understanding its motion. Over the decades following its discovery, careful observations of Uranus's orbit revealed irregularities that could not be explained by the gravitational influence of the known planets. These deviations suggested that Uranus's path was being perturbed by an unseen force—possibly another, yet-undiscovered planet lying farther out.

The idea of a hidden planet was a bold hypothesis, but it gained traction among astronomers in the early 19th century. Two mathematicians, working independently, took on the challenge of calculating the position of this hypothetical planet based on the perturbations in Uranus's orbit. These scientists—**Urbain Le Verrier** in France and **John Couch Adams** in England—approached the problem using similar methods, relying on Newtonian mechanics to predict the gravitational influence of the unknown object.

Le Verrier and Adams tackled a problem of extraordinary complexity. They had to account for the motions of all the known planets, calculate the forces exerted on Uranus by each, and determine the characteristics of the proposed planet that could explain the observed anomalies. Their calculations were feats of intellectual rigor, requiring meticulous attention to detail and a deep understanding of celestial mechanics.

Le Verrier completed his calculations in 1846 and sent his findings to the Berlin Observatory, where astronomer **Johann Galle** received them. On the night of September 23, 1846, Galle and his assistant, Heinrich d'Arrest, turned their telescope to the region of the sky Le Verrier had indicated. Remarkably, they found Neptune almost exactly where Le Verrier's calculations had predicted, just one degree away from the projected location.

This moment was a triumph not only for Le Verrier but for the scientific method itself. It demonstrated that mathematics could be used not merely to describe known phenomena but to predict the existence of entirely new worlds. The discovery of Neptune was a validation of Newtonian mechanics and a testament to the precision and power of theoretical science.

The story of Adams, however, complicates the narrative. Working independently of Le Verrier, Adams had reached similar conclusions about the location of the new planet but struggled to secure the support of British astronomers to confirm his findings. Delays and miscommunications prevented Adams's prediction from being acted upon until after Neptune's discovery in Berlin. This led to a contentious debate over priority and credit, with both Le Verrier and Adams eventually recognized for their contributions.

Neptune's discovery through mathematics rather than direct observation was unprecedented, and it had profound implications for astronomy. It marked the first time a celestial body was located based purely on theoretical predictions, setting a precedent for future discoveries. The method used to find Neptune would later be applied to the search for other planets and phenomena, including Pluto and extrasolar planets orbiting distant stars.

The discovery also highlighted the interconnectedness of the solar system. Neptune's existence was inferred through its gravitational influence on Uranus, a reminder that even in the vast expanse of space, celestial bodies are linked by invisible forces. This principle of interdependence is

Neptune

a cornerstone of modern astrophysics, influencing how scientists study everything from planetary systems to galaxies.

Philosophically, Neptune's discovery represents a shift in humanity's relationship with the cosmos. For millennia, the heavens were understood primarily through direct observation and mythological interpretation. The identification of Neptune, however, demonstrated that the universe could be understood through abstract reasoning and mathematical analysis. It was a moment of demystification and empowerment, showing that the cosmos, while vast and complex, was also knowable.

Neptune's discovery is also a testament to the power of collaboration and perseverance. It involved the work of multiple individuals across different countries, each contributing pieces of a larger puzzle. It underscores the value of shared knowledge and the cumulative progress of science, where the efforts of many build upon one another to achieve breakthroughs.

The story of Neptune's discovery continues to inspire, serving as a reminder that the universe often holds more than meets the eye. It challenges us to look beyond the obvious, to question the anomalies, and to trust in the tools of reason and inquiry. Neptune, a planet revealed not by sight but by calculation, stands as a symbol of human curiosity and the enduring quest to understand the unseen.

As we look outward to the edges of the solar system and beyond, Neptune's discovery offers a guiding principle: that even the faintest hints—a perturbation in an orbit, a flicker of light—can lead to profound revelations. It teaches us that the cosmos is a realm of connections, patterns, and possibilities, waiting to be uncovered by those willing to seek them.

Chapter 6: Voyager's Distant View

The **Voyager 2 spacecraft's** flyby of Neptune in 1989 was a landmark event in the exploration of the outer solar system. It was humanity's first—and, to date, only—close encounter with Neptune, providing a glimpse into a world of stunning complexity, vibrant storms, and enigmatic moons. As the final destination of Voyager 2's grand tour of the outer planets, Neptune offered a fitting conclusion to a mission that had transformed humanity's understanding of the solar system. Yet, even as it answered questions, Voyager 2's fleeting visit left many mysteries unsolved, highlighting both the triumphs and the limitations of robotic exploration.

Voyager 2 approached Neptune after more than a decade of travel through the solar system. Having completed flybys of Jupiter, Saturn, and Uranus, the spacecraft was by then a seasoned explorer, equipped with instruments designed to study planetary atmospheres, magnetic fields, rings, and moons. As it neared Neptune, Voyager 2 began transmitting data and images that revealed the planet in unprecedented detail.

One of the most striking discoveries was Neptune's **dynamic atmosphere**, a realm of extremes that belied the planet's frigid location on the edge of the solar system. Voyager 2 captured images of **the Great Dark Spot**, a massive storm system similar in scale to Jupiter's Great Red Spot. However, unlike its Jovian counterpart, which has persisted for centuries, Neptune's Great Dark Spot appeared transient, disappearing within a few years of its discovery. This impermanence highlighted the dynamic and rapidly changing nature of Neptune's weather.

Neptune's atmosphere also exhibited some of the fastest winds in the solar system, with speeds reaching up to 2,100 kilometers per hour (about 1,300 miles per hour). These winds were detected in the bands of clouds circling the planet, their rapid motion creating patterns of turbulence and shear. The energy driving such extreme winds in a region so distant from the Sun remains a mystery. Scientists suspect that **Neptune's internal heat**, which radiates more energy than the planet receives from the Sun, plays a critical role, fueling convection currents and atmospheric circulation.

Neptune

Voyager 2's observations of Neptune's **magnetic field** added another layer of intrigue. Like Uranus, Neptune's magnetic field was found to be significantly tilted and offset from the planet's center. This unusual configuration created a highly dynamic and asymmetric magnetosphere, with complex interactions between the solar wind and Neptune's magnetic environment. The spacecraft's data hinted at processes occurring deep within Neptune's interior, where conductive materials such as water, ammonia, and methane may create the dynamo effect responsible for the magnetic field.

Neptune's **ring system**, long suspected but not fully understood, was revealed in detail by Voyager 2. The spacecraft confirmed the existence of faint, clumpy rings, with arcs of denser material that defied the expectation of uniformity. These arcs, likely stabilized by the gravitational influence of nearby moons, hinted at ongoing processes that shaped the structure and distribution of ring particles. Voyager's images provided the first direct evidence of these features, transforming the perception of Neptune's rings from theoretical to tangible.

Voyager 2 also turned its instruments toward Neptune's **moons**, capturing detailed images of several, including the enigmatic **Triton**, the largest of Neptune's satellites. Triton was revealed to be a world of startling contrasts, with icy plains, nitrogen geysers, and a retrograde orbit that marked it as a likely captured object from the Kuiper Belt. The spacecraft's data showed active geysers ejecting nitrogen gas into Triton's thin atmosphere, a discovery that hinted at ongoing geological processes and a dynamic interior. Triton's surface, with its smooth plains and fractured terrain, suggested a young and constantly renewing landscape, further fueling speculation about its history and potential habitability.

Despite these achievements, Voyager 2's encounter with Neptune was inherently limited by the nature of a flyby mission. The spacecraft passed within 4,950 kilometers (about 3,000 miles) of Neptune's cloud tops, completing its closest approach in a matter of hours. While this trajectory allowed Voyager to gather an extraordinary amount of data, it also meant that many aspects of Neptune and its system could only be glimpsed briefly, leaving gaps in our understanding.

Neptune

For example, Voyager 2 provided no detailed information about Neptune's deep interior, where processes driving its internal heat and magnetic field remain hidden. Similarly, the fleeting nature of the flyby limited the study of seasonal changes, atmospheric dynamics, and interactions between the rings and moons. Many of Neptune's smaller moons were only faintly detected, their characteristics and origins remaining a mystery.

Voyager 2's data and images, however, were groundbreaking, laying the foundation for all subsequent studies of Neptune. They transformed Neptune from a distant and largely theoretical object into a vivid, dynamic world. The spacecraft's observations continue to inform planetary science, inspiring new hypotheses about the processes that shape ice giants and their systems.

Voyager 2's flyby of Neptune also marked the end of an era. It was the last major planetary encounter of the 20th century, a culmination of the Voyager program's ambitious goal to explore the outer planets. As the spacecraft sped away from Neptune into interstellar space, it left behind a legacy of discovery and a sense of wonder about the solar system's farthest reaches.

Philosophically, Voyager 2's encounter with Neptune embodies the spirit of exploration. It was a mission driven by curiosity and the desire to push beyond known boundaries, to venture into the unknown and bring back knowledge. The spacecraft's journey to Neptune, spanning billions of kilometers and decades of time, is a testament to humanity's ability to transcend its immediate surroundings and seek understanding on a cosmic scale.

The fleeting nature of Voyager 2's visit to Neptune also speaks to the limitations of exploration. While it revealed much, it left even more to discover, highlighting the need for future missions to return to Neptune and study it in greater detail. The spacecraft's flyby was not an ending but a beginning—a first step in unraveling the mysteries of a planet that continues to captivate and challenge.

Neptune, as seen through Voyager 2's eyes, is a world of contrasts: serene and stormy, luminous and shadowed, distant yet intimately connected to the forces that shape the solar system. The spacecraft's

Neptune

distant view reminds us that even the farthest and most elusive worlds are part of a shared cosmic story, one that invites us to continue exploring, questioning, and dreaming.

Chapter 7: Neptune in Myth and Art

Neptune, the furthest gas giant in our solar system, has long been a figure of mystery and imagination, its influence extending far beyond astronomy into the realms of mythology, art, and philosophy. Named for the Roman god of the sea, Neptune embodies the themes of depth, turbulence, and the interplay between beauty and chaos. Its distant, stormy nature and vivid blue appearance have made it a symbol of the unknown and the sublime, inspiring myths, artistic creations, and philosophical reflections that resonate across cultures and eras.

The name Neptune was chosen shortly after the planet's discovery in 1846, aligning with the tradition of naming planets after Roman deities. Neptune, the god of the sea, was a fitting choice for a planet of such depth and mystery, whose azure hue evokes the vastness of Earth's oceans. As the divine ruler of the seas, Neptune was associated with the unpredictable power of water, from its life-giving calm to its destructive fury. These characteristics, mirrored in the storms and winds of the planet itself, capture the essence of Neptune's enigmatic nature.

In Roman mythology, Neptune was not merely a god of water; he was also a symbol of transition and the unseen. The oceans he governed represented both barriers and connections, pathways to unknown lands and metaphors for the depths of human emotion and thought. This duality—of distance and connection, calm and chaos—is reflected in the planet Neptune, a distant sentinel of the solar system that simultaneously beckons us with its beauty and eludes us with its remoteness.

Neptune's mythological roots extend further back to the Greek god Poseidon, a figure of great complexity. Poseidon, like Neptune, was a god of the sea, wielding a trident and commanding both storms and still waters. Yet he was also associated with earthquakes, horses, and the unpredictable forces of nature. Poseidon's volatile temperament, shifting between rage and calm, finds a parallel in Neptune's own dynamic atmosphere, where tranquil blue clouds conceal ferocious winds and massive storms.

Neptune

These mythological connections imbue Neptune with layers of meaning that go beyond its scientific attributes. The planet is not just a celestial body; it is a symbol of depth, mystery, and the boundary between the known and the unknown. It invites us to reflect on the nature of exploration, both physical and intellectual, and to confront the limits of our understanding.

In art, Neptune's vivid blue and stormy character have inspired countless interpretations, from classical paintings and sculptures to modern visual and literary works. Artists of the Renaissance often depicted Neptune or Poseidon as powerful, commanding figures, their tridents raised amidst roaring seas. These images captured the majesty and danger of the oceans, themes that resonate with the scientific reality of the planet Neptune.

In more recent times, Neptune's discovery and scientific exploration have influenced a new wave of creative expression. Science fiction writers and filmmakers have used Neptune as a backdrop for stories of interstellar travel, alien worlds, and the search for meaning in the vastness of space. Its remoteness and the mysteries it holds make it a natural setting for tales of the unknown, a frontier where imagination and reality converge.

Music, too, has drawn inspiration from Neptune. One of the most famous examples is Gustav Holst's orchestral suite *The Planets*, where the movement "Neptune, the Mystic" captures the planet's ethereal and otherworldly qualities. The music, marked by its flowing harmonies and distant, fading echoes, evokes a sense of the infinite, mirroring Neptune's position as the boundary between the solar system and the great unknown beyond.

Neptune's influence extends into philosophy, where its characteristics serve as metaphors for broader concepts. Its distance and isolation have been likened to the limits of human knowledge, the point at which understanding gives way to mystery. Philosophers have drawn parallels between Neptune's hidden depths and the subconscious mind, both vast and largely uncharted. The storms that rage within its atmosphere become symbols of inner turmoil and the forces that shape our thoughts and emotions.

Neptune

In popular culture, Neptune has become a symbol of aspiration and curiosity, representing the human drive to explore and understand the cosmos. Its vivid appearance and dynamic nature remind us that even in the farthest reaches of our solar system, there is beauty and complexity waiting to be discovered. Neptune serves as a reminder that the universe is not merely a collection of objects but a source of inspiration, a canvas for human creativity and wonder.

The story of Neptune in myth and art is a testament to the enduring power of the celestial to inspire humanity. From its origins in ancient mythology to its role in modern science and culture, Neptune bridges the gap between the observable and the imagined, the known and the unknowable. It invites us to dream of distant horizons and to seek meaning in the depths of the universe.

As we continue to explore Neptune through science, its mythological and artistic significance grows richer, reminding us that our understanding of the cosmos is not limited to data and measurements. Neptune, the whispering giant at the edge of the solar system, speaks to us of the infinite possibilities of discovery and the enduring interplay between science, art, and the human spirit.

Chapter 8: The Isolation of Distance

Neptune, the outermost planet in the solar system, occupies a realm defined by distance and remoteness. More than 30 times farther from the Sun than Earth, Neptune exists in a domain where sunlight is faint and the vast emptiness of space dominates. This isolation is not merely a physical characteristic but a profound metaphor, resonating with themes of separation, boundary, and the longing for connection. Neptune's isolation invites reflection not only on the nature of the planet itself but also on humanity's relationship with the cosmos and the challenges of exploring worlds at the edge of the solar system.

Distance shapes every aspect of Neptune. At such a remote location, sunlight takes more than four hours to reach the planet, compared to the eight minutes it takes to reach Earth. The solar energy that bathes Neptune is a mere 1/900th of what warms our planet, plunging the distant giant into perpetual cold. This scarcity of light and heat creates an environment where atmospheric processes, surface dynamics, and even the interactions of moons and rings unfold on vastly different scales than those of the inner planets.

Neptune's isolation also sets it apart in terms of exploration. As the farthest gas giant, it has been visited only once—by the **Voyager 2 spacecraft** in 1989. This fleeting encounter, while transformative, lasted mere hours, providing humanity with a brief glimpse of a planet that remains largely unknown. The distance to Neptune, combined with the challenges of operating spacecraft in such a remote and cold environment, has left much about the planet unexplored. Its atmosphere, interior, and magnetic field, as well as the intricate relationships between its rings and moons, remain subjects of speculation and mystery.

The isolation of Neptune is further underscored by its orbital position. It resides at the outer edge of the solar system's main planetary region, acting as a sentinel between the familiar inner planets and the vast expanse of the Kuiper Belt and the Oort Cloud beyond. This position makes Neptune a boundary in more ways than one: a threshold between the warm, rocky worlds of the inner solar system and the cold, icy bodies of the outer reaches. Neptune's orbit is not merely a path through space; it is a dividing line that separates the known from the unknown.

Philosophically, Neptune's isolation speaks to the human condition. It embodies the tension between longing and separation, the desire to connect with what lies beyond while grappling with the barriers that distance imposes. The planet's remoteness mirrors the challenges of understanding the vastness of the universe, where even our most advanced tools struggle to pierce the darkness and reveal the truths of distant worlds.

Yet, Neptune's isolation is not merely a limitation; it is also an opportunity. Its position at the edge of the solar system allows it to act as a gravitational gatekeeper, influencing the orbits of distant objects and playing a key role in the dynamics of the Kuiper Belt. The planet's strong gravitational field affects the movement of comets and icy bodies, shaping the architecture of the outer solar system and contributing to its stability. Neptune is a silent participant in a cosmic dance that extends far beyond its visible boundaries, connecting it to regions of space that remain largely uncharted.

Neptune's isolation also provides a unique perspective on the nature of planetary systems. Its position and characteristics challenge scientists to consider how planets form and evolve in regions where sunlight is scarce and conditions are extreme. Neptune's dynamic atmosphere, active weather systems, and complex interactions with its moons and rings defy the expectation that remoteness equates to inactivity. Instead, Neptune reveals that even in the farthest reaches of the solar system, processes of creation, destruction, and transformation continue to unfold.

From a cultural perspective, Neptune's isolation has inspired themes of mystery and longing in art, literature, and philosophy. It is a planet that seems to exist on the edge of perception, a distant beacon that calls humanity to imagine and explore. In this way, Neptune becomes a symbol of the unknown, a reminder that even in the face of vast distances, curiosity and creativity can bridge the gap.

The isolation of Neptune is not without its challenges. The physical distance makes exploration daunting, requiring missions that can operate autonomously for decades and survive the extreme cold and

Neptune

darkness of the outer solar system. The time it takes to reach Neptune—at least a decade for a spacecraft launched from Earth—further complicates efforts, demanding advanced propulsion systems, robust technology, and unwavering commitment. These challenges highlight the limits of current technology while also underscoring the potential for innovation and progress.

Despite these difficulties, Neptune's isolation is also a source of its allure. It is a planet that embodies the spirit of exploration, a world that invites us to push the boundaries of what is possible and to seek understanding even in the most remote corners of the cosmos. Neptune's distance is not merely a barrier; it is a challenge that calls humanity to look outward, to imagine what lies beyond, and to embrace the journey of discovery.

In many ways, Neptune's isolation mirrors the philosophical idea of the sublime—the simultaneous experience of awe and fear in the face of something vast and unknowable. The planet's distance and remoteness evoke a sense of wonder, a recognition of the immensity of the universe and our small yet significant place within it. Neptune's storms, its vibrant blue, and its dynamic moons are reminders that even at the edge of the solar system, beauty and complexity persist, defying the desolation that distance might suggest.

To explore Neptune is to confront the isolation of distance, both physically and metaphorically. It is a journey that requires patience, perseverance, and a willingness to embrace the unknown. Yet it is also a journey that promises profound rewards—a deeper understanding of the outer solar system, insights into the processes that govern planetary systems, and a renewed sense of wonder at the vastness of the cosmos.

Neptune's isolation is not an end but a beginning. It marks the boundary of the solar system's traditional planets while pointing toward the uncharted regions beyond. It is a planet that challenges us to look farther, think broader, and dream bigger. Neptune, the sentinel of the outer solar system, is a reminder that the most distant horizons are often the most compelling, calling us to venture beyond the limits of what we know.

Conclusion: The Last Giant

Neptune, the eighth and final planet of the solar system, stands as both an endpoint and a gateway. It is the last of the gas giants, the distant sentinel marking the boundary between the planetary realm we have come to know and the vast, uncharted wilderness beyond. Yet, Neptune is far more than a line in the cosmic sand. It is a world of contradictions, connections, and mysteries that challenge our understanding of the solar system and inspire us to continue our quest for discovery.

As the "last giant," Neptune embodies the culmination of a family of planets that span a remarkable diversity of sizes, compositions, and characteristics. Together, these planets tell the story of a solar system forged by gravitational forces, collisions, and the interplay of light and matter. Neptune's unique traits—its vibrant blue color, dynamic atmosphere, and complex system of moons and rings—make it a fitting conclusion to this narrative. It is a planet that, despite its distance, pulses with energy and activity, defying the expectations of a world so far removed from the Sun's warmth.

Neptune's role as the outermost gas giant also places it in a position of profound influence. Its gravity acts as a gatekeeper for the solar system, shaping the orbits of distant objects in the Kuiper Belt and even affecting the trajectories of comets that pass near the Sun. Neptune's presence ensures a delicate balance in the outer solar system, preventing the chaotic scattering of icy bodies and maintaining the structure of the Kuiper Belt. This gravitational reach extends Neptune's influence far beyond its immediate surroundings, connecting it to regions of the solar system that are only now beginning to be explored.

The planet's vibrant blue hue is a symbol of both beauty and mystery. It hints at the processes that shape its atmosphere—processes we only partially understand. The interplay of methane, sunlight, and other atmospheric components creates Neptune's signature color, while its internal heat drives winds and storms that reach unprecedented speeds. These features challenge our models of planetary atmospheres, pushing us to consider how similar dynamics might operate on exoplanets or in other regions of the cosmos.

Neptune

Neptune's system of rings and moons adds another layer to its story. The faint, clumpy rings, maintained by the gravitational influence of shepherd moons, are a delicate reminder of the transient nature of celestial systems. They are a bridge between the solid bodies of the inner planets and the icy debris fields of the Kuiper Belt. Meanwhile, Neptune's moons, especially the enigmatic Triton, tell tales of capture, collision, and evolution that connect Neptune's history to the broader narrative of the solar system's formation.

Triton, with its retrograde orbit and active geysers, is a microcosm of Neptune's mysteries. It speaks of the past, when the Kuiper Belt was a denser and more dynamic region, and of the present, with its icy plumes and potential subsurface ocean hinting at ongoing geological processes. Triton is both a part of Neptune's system and a link to the outer reaches of the solar system, tying the planet to the frozen worlds that exist beyond its orbit.

The story of Neptune is also a story of exploration. Its discovery in 1846 through the application of mathematics marked a turning point in astronomy, demonstrating the power of reason and observation to reveal the unseen. The Voyager 2 spacecraft's brief but transformative flyby in 1989 provided humanity's first and only close-up views of Neptune, transforming it from a distant dot in telescopes to a dynamic, vibrant world. These milestones underscore the importance of persistence, curiosity, and the tools of science in expanding our understanding of the cosmos.

Yet, for all we have learned, Neptune remains a planet of profound mystery. Its distance and the challenges of exploration mean that many of its secrets are still concealed. What drives the planet's extreme winds? How does its magnetic field interact with the solar wind and its moons? What lies beneath its cloud tops, in the deep layers where internal heat and pressure create exotic states of matter? These questions remind us that Neptune is not a conclusion but a frontier—a place where the known and the unknown meet.

Philosophically, Neptune represents the tension between isolation and connection. It is the farthest planet, existing in a realm of cold and darkness, yet it is intrinsically linked to the rest of the solar system through gravity, history, and the shared processes of planetary evolution.

Neptune

Its position as the "last giant" makes it both a boundary and a bridge, challenging us to expand our horizons and consider what lies beyond.

Neptune also speaks to the resilience of nature. Despite its distance from the Sun, it is a world of energy and activity, with storms that rival those of Jupiter and a system of moons and rings that are dynamic and evolving. It reminds us that even in the most remote corners of the solar system, life—if not in a biological sense, then in a geological and atmospheric sense—thrives.

As we look to the future, Neptune stands as a beacon for exploration. The data from Voyager 2, while invaluable, is only the beginning. A dedicated mission to Neptune, equipped with modern instruments and the ability to linger in orbit, could unlock the planet's secrets and provide insights into the nature of ice giants, both in our solar system and beyond. Such a mission would not only deepen our understanding of Neptune but also inform the study of exoplanets, many of which share characteristics with this distant giant.

In the end, Neptune is more than a planet; it is a symbol of the human spirit. Its vibrant blue hue, stormy atmosphere, and distant position remind us that the universe is vast and full of wonders waiting to be discovered. It challenges us to look outward, to question, and to imagine, embodying the enduring drive to explore and understand.

Neptune, the last giant, is a threshold to the infinite. It is the culmination of one journey and the beginning of countless others—a reminder that the most distant horizons are often the most profound.

Neptune

End Note: The Whisper of the Kuiper Belt

As the outermost planet in the solar system, Neptune stands as a gateway to the Kuiper Belt, a vast and mysterious region of icy bodies that stretches beyond its orbit. This belt, a relic of the solar system's formation, is populated by objects ranging from small chunks of rock and ice to dwarf planets like Pluto. It is a realm of primordial material, untouched by the heat and gravitational influences that have shaped the inner planets. In many ways, the Kuiper Belt is a whisper from the past, a reminder of the solar system's origins and an invitation to explore the unknown.

Neptune's gravitational influence plays a crucial role in shaping the Kuiper Belt. The planet's immense gravity shepherds and scatters objects in this region, creating a structure marked by resonances and gaps. For example, many Kuiper Belt objects (KBOs) are in orbital resonance with Neptune, meaning their orbits are synchronized with the planet's. Pluto, the most famous of these objects, follows a 2:3 resonance, completing two orbits of the Sun for every three orbits of Neptune. This gravitational dance stabilizes Pluto's orbit and prevents it from colliding with Neptune, highlighting the interconnected dynamics of the outer solar system.

Beyond resonances, Neptune's gravity also acts as a gatekeeper, influencing the trajectories of objects that venture closer to the inner solar system. Many comets, originating from the Kuiper Belt, pass through Neptune's domain before making their way toward the Sun. In this role, Neptune serves as a gravitational buffer, deflecting or capturing some objects while allowing others to continue their journeys. This interplay shapes not only the Kuiper Belt but also the population of comets that we observe from Earth, linking Neptune to the broader processes that govern the solar system.

The Kuiper Belt itself is a region of remarkable diversity and significance. It contains some of the most ancient and pristine material in the solar system, offering a window into the conditions that existed during its formation. Studying KBOs allows scientists to piece together the story of how the planets formed and how the solar system evolved over billions of years. These objects are like cosmic time capsules, preserving the chemical and physical properties of the early solar nebula.

Neptune's moon Triton, with its retrograde orbit and icy surface, is thought to be a captured KBO, providing a tangible connection between the planet and the belt. Triton's characteristics—its nitrogen geysers, cryovolcanic plains, and potential subsurface ocean—offer a glimpse of what might be found among the Kuiper Belt's more massive bodies. Triton serves as a proxy for exploring the Kuiper Belt, a reminder that even objects close to Neptune are shaped by the forces and histories of this distant region.

The Kuiper Belt's significance extends beyond the solar system. Many of the processes observed in this region, such as the interactions between Neptune and KBOs or the chemical composition of icy bodies, are likely to occur in other planetary systems. By studying the Kuiper Belt, scientists gain insights into the dynamics of exoplanetary systems, where distant planets and debris belts may interact in similar ways. The lessons learned from Neptune and the Kuiper Belt thus have implications for understanding the broader universe.

Philosophically, the Kuiper Belt is a frontier in every sense of the word. It represents the outer boundary of the solar system as we know it, a place where the familiar fades into the unknown. Yet it also serves as a bridge, connecting the planets to the vast interstellar spaces beyond. Neptune, as the last giant, is both a guardian of the inner solar system and a sentinel of the Kuiper Belt, marking the transition between two realms.

The Kuiper Belt also challenges us to think about the nature of boundaries and exploration. It is not a single, defined region but a gradient, extending outward into the even more distant Oort Cloud and ultimately into interstellar space. Its whisper is one of continuity, a reminder that the solar system is not a closed system but part of a larger cosmic network.

Exploring the Kuiper Belt and its connections to Neptune is a daunting but essential task. Missions like NASA's **New Horizons**, which flew past Pluto and the KBO Arrokoth, have provided invaluable glimpses into this distant region, but there is much more to discover. A dedicated mission

Neptune

to Neptune, equipped with modern technology, could not only study the planet in greater detail but also serve as a staging ground for exploring the Kuiper Belt. Such a mission could examine the interactions between Neptune's gravity and KBOs, capture detailed images of these icy bodies, and investigate their compositions and histories.

Neptune's position as a gateway to the Kuiper Belt underscores its importance in the story of the solar system. It is not merely a planet but a link to the uncharted realms beyond—a bridge between the past and the future, the known and the unknown. The Kuiper Belt, in turn, is a whisper of what lies beyond Neptune, a region that calls to us with the promise of discovery and the allure of mystery.

In the end, Neptune and the Kuiper Belt remind us of the infinite possibilities of exploration. They challenge us to look beyond the immediate, to venture into the vastness of space, and to embrace the unknown with curiosity and wonder. The whisper of the Kuiper Belt is a call to continue our journey, to push the boundaries of what we know, and to dream of what lies beyond.

Neptune, the last giant, is not an end but a beginning—a threshold to the infinite, where the whispers of the Kuiper Belt beckon us to explore further, to imagine more, and to never stop seeking answers to the mysteries of the universe.

www.ingramcontent.com/pod-product-compliance
Lightning Source LLC
Chambersburg PA
CBHW070944220526
45469CB00007B/2507